Herbicide Resistance And Phosphatase Enzyme Activity Of Phosphate Solubilizing Bacteria In Rhizosphere Soil

by

Murugan Karuppasamy
PG Research Scholar, PG Department of Microbiology, Sri Paramakalyani College, Alwarkurichi - 627 412, Tenkasi District, Tamilnadu, India.
*Corresponding author telephone: 94425 06790,
e-mail: micromurugan2013@gmail.com

Double 9 Books

Herbicide Resistance And Phosphatase Enzyme Activity Of Phosphate Solubilizing Bacteria In Rhizosphere Soil

by Murugan Karuppasamy

ISBN: 978-93-63057-91-3

Published by

DOUBLE 9 BOOKS

2/13-B, Ansari Road
Daryaganj, New Delhi – 110002
info@double9books.com
www.double9books.com
Tel. 011-40042856

ABOUT THE AUTHOR

Murugan Karuppasamy is already worked as an ex Assistant Professor & Head and Microbiologist & Quality Controller in the discipline of Microbiology at college and industry. He completed his degree in M.Phil. in Microbiology under Life Science, in 2011 at Manonmaniam Sundaranar University of Tirunelveli, India. He did his M.Sc. in Microbiology, in 2002 at Sri Paramakalyani College of Alwarkurichi, India. He has also published an original research article in an international journal. His areas of research interests are all the branches of Microbiology.

CONTENTS

ACKNOWLEDGEMENT

First and foremost of all, I thank the almighty for blessing me to complete my course Master of Science in Microbiology and the project work successfully.

I am grateful to express my deep sense of gratitude to my profound research guide **Dr. Gowrishankar Rajam, M.Sc., Ph.D.** I owe him personnel indebtedness for giving valuable suggestion for problem, his guidance, wise council, unstinted encouragement both psychologically and academically. His keen interest and effort helped me in completing this project. I feel pleasure, immense pride and as most blessed person in presenting the work under his guidance.

I must pay respect and gratefulness to **Dr. R. Palaniappan, M.Sc., Ph.D.** for allowing me to have free access of lab and his timely advice.

I am thankful to **Dr. S. Ramesh, M.Sc., Ph.D.** and **Mrs. V. Jeyanthi Kumari, M.Sc.** for timely suggestions they gave me during the period of my work.

My special thanks to secretary, professor **K. H. Narayanan**, the principal **Dr. G. Devarajan, M.Sc., Ph.D.** and the management of our institutions for providing all the necessary facilities.

I am thankful to **Mr. CT. Kannan** and associate, at **ABHI DTP**, Tirunelveli junction for the efforts in bringing this book to the best of my satisfaction by printing in a delicate manner.

Finally, I have been bestowed with the privilege of expressing my gratitude to my parents, brothers and other well wishers for their moral support, blessing and prayers.

Murugan Karuppasamy

ABSTRACT

Phosphate solubilizing microbes plays an important role in plant nutrition through increase in phosphate uptake by plants and used as biofertilizers of agricultural crops. Phosphate is one of the most vital macronutrient required for the growth and development of plants. A large number of microorganisms present in the rhizosphere are known to solubilizer and make available the insoluble phosphorus in the available form to the plants. Plant growth promoting rhizobacteria (PGPR) mediate the soil processes such as decomposition, nutrient mobilization, mineralization, solubilization, nitrogen fixation and growth hormone production. Rhizosphere soil of *Oryza sativa* (paddy plant) was collected from cultivated crop field and enumerated for total heterotrophic and phosphate solubilizing bacterial population. In this study, twenty five strains of phosphate solubilizing bacteria (PSB) were isolated by plate assay and characterized biochemically. Mineral phosphate solubilizing (MPS) activities of all isolates were tested on tricalcium phosphate of Pikovskaya's agar medium by analyzing the soluble phosphate content after 24 hours of incubation at 37°C. The aim of the present study was to screen the phosphatase enzymes of phosphate solubilizing bacteria and to access their potential tolerance to herbicides (one shot). The heterotrophic bacterial load was found in the order of 10 CFU/gm and the phosphate solubilizer of 10 CFU/gm. The phosphate solubilizing bacteria was composed of *Bacillus* sp. and *Micrococcus* sp. with a lone of *Pseudomonas* sp. Both *Bacillus* sp. and *Micrococcus* sp. were noted to produce phosphatase at neutral pH and at 45°C with the exception of *Bacillus* sp. phosphate solubilizer (PS) 25. Probably this bacterium is secreting alkaline phosphatase that has temperature optima at 35°C. None of the three representative bacterial strains were found to tolerate the most commonly used herbicide (one shot) at the concentration of 20 ppm. Only *Pseudomonas* sp. was recorded to tolerate this herbicide at 10 ppm. Further understandings on the nature of the enzyme, optimization on its activity would provide basic data for the commercial production of these enzymes.

KEYWORDS: Rhizosphere soil, Pikovskaya's agar medium, Griseofulvin, Phosphatase enzyme, Phenolphthalein diphosphate and Herbicide.

1
INTRODUCTION

Soil microorganisms have great potential in providing soil phosphates for plant growth. Phosphorus biofertilizers can help to increase the accessibility of accumulated phosphates for plant growth by solubilization (Goldstein, 1986; Gyaneshwar *et al.*, 1998). Rhizosphere is the soil zone surrounding the plant roots while rhizoplane is directly in contact with the roots (Kennedy, 2005). Plant roots exuded the organic contents in the rhizosphere and subsequently increased the microbial activity and termed as "rhizosphere effect" by Hiltner (1904). Microorganisms exerted beneficial effect on plant growth and development through different means is termed as plant growth promoting rhizobacteria (PGPR) (Vessey, 2003). Plant growth promoting rhizobacteria accounts for about 2-5% of total the rhizobacteria involved in plant growth promotion (Antoun and Kloepper, 2001).

Phosphate solubilizing microorganisms (PSM) have attracted the researchers to exploit their potential to utilize phosphate reserves in semi-arid regions and to enhance the crop yields (Goldstein *et al.*, 1993; Fasim *et al.*, 2002; Khan *et al.*, 2007). Phosphate solubilizing microorganisms have established their role for optimum growth of plants under nutrient imbalance conditions (Glick, 1995; Iguala *et al.*, 2001; Wu *et al.*, 2005).

Phosphorus (P) is the second most important plant nutrient after nitrogen (Donahue *et al.*, 1990). Phosphorus is an essential macronutrient for growth and development of plants involved in important metabolic pathways like photosynthesis, biological oxidation, nutrient uptake and cell division (Illmer and Schinner, 1992). Worldwide soils are supplemented with inorganic phosphate as chemical fertilizers to support crop production but repeated use of fertilizers deteriorates soil quality (Tewari *et al.*, 2004).

The problem of phosphate fertilization may become serious in the coming years because of the fact that manufacture of phosphatic fertilizers requires the use of non-renewable resources such as high grade rock phosphate which are getting depleted progressively and becoming costlier. It is estimated that about 98% of Indian soils contain insufficient amounts of available phosphorous to support maximum plant growth. So to increase the soil fertility and agricultural productivity phosphatic fertilizers are widely used.

Rock phosphates are the only commercial raw material for the production of phosphatic fertilizers. The total world reserves of rock phosphate are estimated to be about 27,000 million tones, of which 80% are located in USA, USSR and Morocco. In India, rock phosphate deposits are of low grade (contain less than 25% of the P_2O_5) and do not satisfy the requisite specification of fertilizer industry. Beneficiation of such low grade rock phosphates is successful only in acidic soils and in neutral and an alkaline soils, rock phosphate turns out to be rather poor source of phosphorus (Neelima Garg *et al.*, 1989).

Many soil microorganisms including bacteria, fungi, yeast and actinomycetes are capable of solubilizing the insoluble inorganic phosphates (Skujins, 1967). Soil microorganisms play a very significant role in mobilizing phosphorous for the use of plants in bringing about changes in the pH of the soil microenvironment leading to the solubilization of native as well as added insoluble phosphorous.

Sen and Paul (1957) observed the solubilization of calcium and iron phosphate in liquid medium by *Bacillus subtilis, Bacillus megaterium* and *Bacillus mesentericus* with ammonium sulphate as the nitrogen source. They also isolated the bacteria capable of solubilizing tricalcium phosphate from the glands of *Cassia occidentals*.

The soil environment surrounding the plant roots is the zone of intense microbial activity. Swaby and Sperber (1959) reported that population of phosphate dissolving microorganisms are more in the rhizosphere (20-40% of the total population) compared to non-rhizosphere (10-15% of the total population). Solubilization of phosphorous by phosphate solubilizing microorganisms (PSM) is attributed to excretion of organic acids (Sperber, 1957).

Artidave and Patel (1999) worked on inorganic phosphate solubilizing soil Pseudomonads and isolated thirty eight *Pseudomonas* sp. which were examined for their tricalcium phosphate solubilizing activity in Pikovskaya's broth for twenty one days? Though all cultures solubilizer tricalcium phosphate in liquid medium only thirteen cultures showed zone of phosphate dissolution on Pikovskaya's agar medium. The decomposition and mineralization of organic phosphate occurs under the influence of specific enzymes of phosphatase. Hence, it has an essential role to play in the phosphorous cycle in the environment.

Chemical pesticides are commonly used by farmers and corporate in agriculture to protect the crops, but these activities are depleting the soil fertility and soil health. The use of chemical pesticides in Indian agriculture drastically increased in recent years. The word pesticides include a heterogeneous group of chemicals developed to control a variety of pests. Pesticides are generally categorized as insecticides, herbicides and fungicides according to the type of pest which they have shown efficacious action (Miligi *et al.*, 2006).

As microorganisms play an important role in many soil biological processes, including nitrogen transformations, organic matter decomposition, nutrient release and their availability, as well as stabilize the soil structure and affect its fertility, investigated by (Vyas, 1988; Edwards and Bater, 1990; Khan and Sculion, 2000). However, soil is the most important site of biological interactions. The indiscriminate use of pesticides disturbs the soil environment by affecting flora and fauna including soil microbial flora, soil enzymes and also the physicochemical properties of soil like pH, salinity, alkalinity leading to infertility of soil. The important micro flora, beneficial for the growth of plants includes nitrogen fixing bacteria and phosphate solubilizing bacteria, present in the rhizosphere of the plant.

Various agrochemicals e.g. herbicides when applied intensively and erratically on herbicide resistant non-transgenic and transgenic crops to control the noxious weeds leads to their accumulation in soil to a dangerous level that affects growth, survival, efficiency and quality of beneficial microbial communities of soil (Srinivas *et al.,* 2008; Pereira *et al.,* 2008; Ahemad *et al.,* 2009). The naturally abundant plant growth promoting rhizobacteria are also metabolically inactivated through the uptake of herbicides applied in excess to the soil (Barriuso *et al.,* 2010). In contrast, a few microorganisms found to be tolerant or resistant towards specific herbicides. It is of great concern that how to reduce the effect of herbicides on the beneficial microorganisms and at the same time it is of great interest to screen out microorganisms which are tolerant / resistant to herbicides. Herbicides are applied in modern agricultural practices to offset the plant growth restricting weeds and subsequently to augment the productivity of crops (Ahmed *et al.,* 2009). The intensive application of herbicides leads to their accumulation in soils to a level that adversely affects both the quality and biological composition of soils (Srinivas *et al.,* 2008), the naturally abundant plant growth promoting rhizobacteria (PGPR) are metabolically inactivated by taking up excessive herbicides (Ahemad and Khan, 2010; Bellinaso *et al.,* 2003).

Many pesticides showed no detectable effects on soil microorganisms at the recommended application rates. However, application at increased rates is often reported as most of the farmers decide based on their own experience of the effective pest control. Repeated and overuse of pesticides in agriculture is a matter of concern because these chemicals are recognized as a source of potential adverse impacts on the metabolic activities of soil microorganisms as well as their plant growth promoting characteristics (Wani *et al.,* 2005; Ahemad *et al.,* 2009). Some microorganisms (called as phosphate solubilizing microorganisms - PSMs) perform phosphate solubilization. Their growth and phosphate solubilizing activity may also be affected by the pesticides leading to the imbalance phosphorous nutrition for the crop plants. However, the effect of pesticides on microbial growth and their activity especially phosphate solubilization can only be assessed using microorganisms which are tolerant to the pesticide of concern (Oves *et al.,* 2009).

2
SCOPE AND AIM OF THE STUDY

The present study falls on the following lines:

2.1 Bacteriological Examination

- To study, the enumeration of total heterotrophic bacterial population (THBP) in rhizosphere soil sample by total viable count.
- To study, the isolation and enumeration of phosphate solubilizing bacteria (PSB) from the rhizosphere soil sample by spread plate technique.
- To maintain the phosphate solubilizing bacterial isolates on nutrient agar slants.
- To identify, the phosphate solubilizing bacterial isolates using the scheme of Aiso and Simidu (1962).

2.2 Experimental Works

- To observe, the determination of phosphatase enzyme activity in phosphate solubilizing bacteria
    - Preparation of the soup of phosphatase enzyme source
    - Phosphatase enzyme assay by spectrophotometric method (Spectronic 20D)
- To study, the culture condition optimization for phosphatase enzyme production
    - Effect of pH on phosphatase enzyme production
    - Effect of temperature on phosphatase enzyme production
- To evaluate, the herbicide resistance in phosphate solubilizer by plate assay method

3
REVIEW OF LITERATURE

In recent years, attention has been paid to the microbial products, which have significant agricultural value. Microorganisms produce secondary metabolites and some of which have phosphate solubilizing capabilities. Some of the microorganisms are isolated from the soil and their phosphate solubilization capabilities were improved by using physical and chemical mutagenesis.

Phosphate Solubilization

Soil generally contains adequate amount of organic and inorganic phosphorus. But most of these remain unavailable to plants. The efficiency of water soluble phosphorus is usually low and its recovery does not exceed 20% (Darmwall *et al.*, 1989). But by the help of phosphate solubilizers, phosphate availability in soil and the utilization of phosphate by plants are increased. (Gaur and Sacher, 1980). Microbial solubilization of insoluble phosphate, especially low grade and its use in agriculture is receiving greater attention. This process not only compensates for higher cost of manufacturing fertilizers in industry but also mobilizes the fertilizers added to the soil.

Ahamed and Jha (1968) isolated phosphate solubilizing bacteria, fungi and actinomyces from soil samples of Bihar on carrot extract agar. From their studies, it is identified that bacteria were more in number than other types of microbes whereas, the fungi performed better solubilization than other types of microbes. Bhurat and Sen (1968) examined the leaf surface of wheat, pea, barley and gram for the presence of phosphate solubilizing bacteria and isolated nearly fifteen isolates of various species.

Thakker *et al.,* (1993) isolated phosphate solubilizing Gram-positive and Gram-negative bacteria from different crops and compost samples and identified the pure culture on the basis of cell morphology, staining reaction, biochemical and physiological characteristics as outlined in 8th edition of Bergey's Manual of Determinative and Systemic Bacteriology.

Katznelson and Bose (1959) isolated rhizosphere bacteria from wheat, that oxidized glucose and alanine and they observed over one third of cultures tested were capable of dissolving insoluble phosphate in the form of calcium phosphate. Goswami and Sen (1962) isolated three bacterial strains and examined for their tri calcium phosphate solubilizing activity in Pikovskaya's broth.

Kucey (1983) phosphate solubilizing, total bacterial and fungal populations were determined by serial dilution and plate counting. He reported that the fungi were superior to bacteria in solubilizing both freshly precipitated calcium phosphate and Idaho rock phosphate. Fungi also retained this ability over many sub culturing transfers. A high percentage of the bacterial isolates lost their solubilizing ability when sub cultured. A significant correlation was found between an organism's ability to solubilize freshly precipitated calcium phosphate in agar plates and Idaho rock phosphate in solution culture.

Cunningham and Kuiack (1992) isolated *Penicillium bilaii* to study the solubilization of mineral phosphates and enhance plant uptake of phosphate. Using agar media with calcium phosphate and the pH indicator alizarin red S, the influence of the medium composition on phosphate solubility and medium acidification was recorded. Three gram-positive and four gram-negative phosphate solubilizing bacteria were isolated from rhizosphere of different crops and compost samples. Among the bacteria, *Enterobacter aerogens,* was found to be most effective tri calcium phosphate (TCP) and rock phosphate solubilizer. In presence of tri calcium phosphate and rock phosphate, the effect of carbon, nitrogen sources and pH of the medium on phosphate solubilization were also studied (Thakkar *et al.*, 1993).

Two species like *Penicillium aurantiogriseum* and *Pseudomonas* sp. having high abilities in solubilizing inorganic phosphates (hydroxyl apatite and brushite) were used to examine solubilization mechanisms. No direct contact between microorganisms and calcium phosphates (Ca-Ps) were necessary for effective solubilization (Illmer and Schinner, 1995).

Sudhansu S Pal (1999) isolated twenty three phosphate solubilizing bacterial cultures. Of the twenty three isolates the acid tolerant strain was identified as *Bacillus* sp. on the basis of morphological and biochemical tests as described in bacteriological manual. Artidave and Patel (1999) isolated phosphate solubilizing bacteria from soil. They identified the bacteria up to genus level by positive test of the organism for pigment production on King's B medium and by cultural, morphological and biochemical characteristics. Based on the above results and oxidase positive reaction in King's B medium, they identified the isolated culture was *Pseudomonades.*

A bacterial strain isolated from rhizospheric soil of grasses growing spontaneously in Spanish soil, actively solubilized phosphates *in vitro* when calcium phosphate was used as phosphorus source. This strain was Gram-negative, strictly aerobic, rod-shaped and motile. The strain produced catalase, but not oxidase. Cellulose, casein, starch, gelatin, aesculin and urea were not hydrolyzed. Growth was observed with many carbohydrates as the carbon source. (Peix *et al.*, 2003).

A phosphate solubilizing microorganism isolated from rhizospheric soil and temporarily identified as *Burkholderiaglathei* produced gluconate and acetate using glucose as a carbon source and its metabolic activity caused the pH of the liquid medium to decrease as low as 4.4. Out of several hundred colonies that grew on Pikovskaya's medium thirteen best isolates were selected based on the solubilization of insoluble phosphates in liquid culture and further characterized and identified (Chung *et al.*, 2005).

Fankem *et al.*, (2006) collected seven rhizosphere soil samples from oilpalm tree of Cameroon. The collected soil samples were air dried, crushed to pass through 2mm sieves and the phosphate solubilizing bacteria were isolated by using the soil samples. For the estimation of phosphate solubilization Reyas basal medium having calcium, aluminium and iron phosphates was used. At the end of incubation time, it appeared that, phosphate solubilization resulted from a combined effect of pH decrease of the media and organic acids production. Furthermore, each of the tested isolates was able to produce at least one of the most important organic acids such as citrate, malate and tartrate. Among the ten isolates tested, three were identified as *Pseudomonas fluorescens* and would be considered as potential biofertilizers.

A field experiment has been conducted with four systemic herbicides at their recommended field rates to investigate their effects on growth and activities of aerobic, non-symbiotic, nitrogen fixing bacteria and phosphate solubilizing microorganisms in relation to availability of nitrogen and phosphorus in the rhizosphere soils as well as yield of the rice crop. Application of herbicides, in general, highly stimulated the population and activities of the target microorganisms, which resulted in a greater amount of atmospheric nitrogen fixation and phosphate solubilization in the rhizosphere soils of the test crop. (Das and Debnath, 2006).

Narula *et al.*, (2007) isolated phosphate solubilizing bacteria from the rhizosphere soil of eternal rays and reference strains of *A. chroococcum, Pantoea, Pseudomonas* were compared for phosphate solubilizing ability over a period of eighteen days. Without much effect on pH, cumulative effect of these entire factors might have had an indirect effect on phosphate solubilization.

Shahab and Ahmed (2008) studied that the phosphate solubilization efficiency of ten soil bacteria for various parameters like carbon sources such as glucose, fructose, sucrose and lactose, variable concentration of sodium chloride and glucose. Glucose was the most favorable carbon source for solubilization while lactose is the least favorable carbon source. *Acinetobacter iwoffi, Pseudomonas aeruginosa and Bacillus thuringiensis* were found to be the most promising isolates.

Joseph and Jisha (2009) isolated phosphate solubilizing bacteria possessing the ability to solubilize insoluble inorganic phosphates from rhizosphere soil. Eighty one potential phosphate solubilizing bacteria thus obtained were quantitatively screened for phosphate solubilization. Of these, four bacteria such as *Acetobacter liquefaciens, Acetobacter* sp., *Pseudomonas gladioli* and one unidentified strain found to be efficient phosphate solubilizers. They were selected for further evaluation, and found that they solubilize tricalcium phosphate in buffered as well as non-buffered media. The efficiency of phosphate solubilization was decreased in buffered media compared to non-buffered media. The buffering capacity of the medium reduced the effectiveness of phosphate solubilizing bacteria in releasing phosphate from tricalcium phosphates.

Kumar *et al.*, (2010) isolated six phosphate solubilizing bacteria from paddy fields of Eastern Uttar Pradesh, India and identified as members of *Enterobacter* and

Exiguobacterium genera. Of the six isolates, *Enterobacter* sp. exhibited high level of phosphate solubilization in liquid medium. *Exiguobacterium* sp. showed increased phosphate solubilization efficiency under alkaline pH.

A simple method of assaying soil phosphatase activity was described. It involves colorimetric estimation of the p-nitrophenol released by phosphatase activity when soil is incubated with buffered (pH 6.5) sodium p-nitrophenyl phosphate solution and toluene at 37° C for one hour. The method is rapid and precise, and it has significant advantages over methods previously proposed for assay of soil phosphatase activity (Tabatabai and Bremner. 1969).

The occurrence and distribution of a special group of bacteria, capable of dissolving insoluble phosphates, were studied in marine environments, especially in sediments. The phosphatase activity was also investigated. There was a positive correlation between the total phosphate content and the phosphatase activity. The phosphatase activity was recorded in all samples, irrespective of salinity variations (Ayyakkannu and Chandramohan, 1971).

The bacteria *Lysobacterenzymogenes* was studied for its extracellular phosphatase during the stationary phase of growth. The cells also produced a cell-associated alkaline phosphatase. This enzyme is found in the particulate fraction of cell extracts and may be membrane bound. The production of both phosphatases, especially the extracellular enzyme, is reduced by inorganic phosphate (Richard and Tigerstrom, 1984).

Bacterial strains were obtained from a soil sample and culture collection center and they were used for enzyme production in a low-concentration phosphate medium. Various concentrations of sodium phosphate or calcium were added in experiments to study the regulation of phosphatase synthesis. Alkaline phosphatase activity was measured spectrophotometrically by monitoring the release of para nitro phenol from para nitro phenyl phosphate (PNPP) at 400 nm (Prada *et al.*, 1996).

The use of phosphate solubilizing bacteria as inoculants simultaneously increases phosphate uptake by the plant and crop yield. Strains from the genera *Pseudomonas, Bacillus* and *Rhizobium* are among the most powerful phosphate solubilizers. The principal mechanism for mineral phosphate solubilization is the production of organic acids, and acid phosphatase play a major role in the mineralization of organic phosphorous in soil. Several phosphatase encoding genes have been cloned and characterized and a few genes involved in mineral phosphate solubilization have been isolated. Therefore, genetic manipulation of phosphate solubilizing bacteria to improve their ability to improve plant growth may include cloning genes involved in both mineral and organic phosphate solubilization, followed by their expression in selected rhizobacterial strains (Rodriguez and Fraga, 1999).

The strains, viz., phosphobacteria, yeast and mildew were compared for their phosphorus dissolving mechanism and phosphorus solubilizing abilities. Acids, acid phosphatase, and alkaline phosphatase which have synergic effects on phosphate dissolution were studied. The variation in activities of acid phosphatase and alkaline phosphatase of phosphate solubilizing microbes were also examined (ChuanQing and WeiYi, 2005).

Ponmurugan and Gopi (2006) conducted an experiment to enumerate the population density of phosphobacteria in the rhizosphere soils of various crops using Ketznelson and Bose medium following dilution plate technique. Efforts have been made to isolate phosphobacteria from these soils and isolated strains were inoculated in specific media containing specific substrates to produce growth regulating substances such as indole acetic acid and gibberellic acid and phosphatase enzyme.

4
MATERIALS AND METHODS

4.1 Sampling Area

The sampling area is a crop field in the surroundings of Alwarkurichi village, Tirunelveli Kattabomman district. Here, the surroundings of this area are completely covered by crop field. Most of the people in this village are farmers. They cultivated various types of crops and vegetable plants. The crop field is enriched by applying phosphatic fertilizers and natural phosphate containing manures. So the plants rhizosphere root regions can be expected to load of phosphate solubilizing bacterial flora. Some pests are affected the rhizosphere root region of plants and to retard the growth of plants and to decent the crop yield. So pesticides are very important to needed for the control of insect pests.

4.2 Sampling

Rhizosphere acidic soil samples of *Oryza sativa* (paddy plant) was collected from cultivated crop field. The acidic soil samples were collected from various corners and middle regions of the paddy field very important. Because, the content of microbial load is variable from region to region of the paddy field. Where, good growth of paddy plant absorbs high amount of soluble and insoluble phosphates around their rhizosphere root regions. Hence, their rhizoplane of paddy plant root regions would have appreciable concentrations of phosphate solubilizing bacterial flora. So, these selected paddy plant root regions were chosen for the isolation of phosphate solubilizing bacterial flora. The rhizoplane root regions of the paddy plants having higher density of phosphate solubilizing bacterial species and the phosphate solubilizing capacity were highly greater.

During acidic soil sample collection from the paddy crop field, the following care was taken.

1. The rhizosphere acidic soil sample of paddy field were taken and transferred into the sterile conical flask with the help of sterile knife. The acidic soil samples from rhizosphere root regions of the paddy field were taken compulsorily.
2. The sterile conical flask was transferred in an ice pack (>4°C). Because the activity of bacterial flora would not be altered.
3. Then, the soil sample was brought to the laboratory subjected to microbial analysis within six hours to prevent the reduction of bacterial countings.

4.3 Bacteriological Examination

Selective culture media employed for the isolation of phosphate solubilizing bacteria

1. Pikovskaya's agar medium (modified by Sundara Rao and Sinha, 1963).
2. Hydroxy apatite agar medium (modified by Ayyakkannu and Chandramohan, 1970).
3. Soil extract agar medium (modified by Katznelson and Bose, 1959).

4.4 Enumeration of Total Heterotrophic Bacterial Population (THBP) in Rhizosphere Soil Sample by Total Viable Count

The collected soil sample was allowed to attain room temperature. The soil sample is used to determine the total viable count and the density of phosphate solubilizing bacterial flora.

Steps in serial dilution technique

About ten gm of soil sample was taken from the conical flask and aseptically transferred into a conical flask containing 90 ml of sterile distilled water. It represents 10^{-1} dilution. One ml was pipette out from the conical flask and transferred into a test tube containing 9 ml of sterile distilled water. It represents 10^{-2} dilution. This is repeated up to 10^{-5} dilution. Care was taken to mix the sample solution of each dilution thoroughly in a vortex mixture prior to pipetting out.

Steps in pour plate technique

Petri dishes and pipettes were sterilized in hot-air oven (180°C - 2 hours). Then, the nutrient agar medium was prepared, autoclaved and cooled to 45°C. 15 μl stock solution of antifungal agent griseofulvin (1 mg / 1 ml) was added aseptically to the medium to suppress the growth of fungal flora. From the above serially diluted sample, one ml of three consecutive dilutions to be tested was pipette out on to the sterile Petri dishes in replicates. Then the cooled nutrient agar medium was added to the plates (approximately 20 ml) and mixed thoroughly with the inoculums by rotating the plate in clockwise and counterclockwise directions, taking care not to spill the medium. Care was taken to see that while mixing the nutrient agar medium and the inoculums, the media does not get solidify while rotating it clockwise or counter clockwise directions. The pour plated materials after solidification were inverted and incubated at 37°C for a period of 24-48 hours. After incubation, plates with colonies ranging from 30 to 300 were selected for counting.

Isolation and enumeration of phosphate solubilizing bacteria (PSB) from the rhizosphere soil sample by spread plate technique

About ten gm of soil sample was weighed aseptically and dissolved in 90 ml of sterile distilled water and mixed thoroughly in a rotator shaker for the maximum recovery of bacterial cells from the soil. This represents 10-1 dilution. From the above dilution, the sample was diluted up to 10-5 dilution. Then Pikovskaya's agar medium, hydroxy apatite agar medium and soil extract agar medium were prepared, sterilized and cooled, to this 15µl stock solution of griseofulvin (1 mg / 1 ml), antifungal agent was added aseptically and dispensed the medium in sterile Petri dishes. From the above 10^{-3}, 10^{-4} and 10^{-5} dilutions 0.1 ml of the sample was pipette out and inoculated on the air dried sterile Pikovskaya's agar medium, hydroxy apatite agar medium and soil extract agar medium, these were spread plated. The replicates and control were maintained. One set of plates were incubated at 37°C for 24 hours in an incubator and the another set of plates were incubated at room temperature for 48 hours. After sufficient incubation, the plates were observed for the presence of phosphate solubilizer on the basis of transparent clear zone formation surrounding the colony.

4.5 Maintenance of Phosphate Solubilizing Bacterial Isolates

Isolations were made at random from Pikovskaya's agar medium or other selective media plates containing countable number of colonies (30-300 colonies). After recording the morphological characteristics of the colony and pigmentation, they were checked for their purity by repeated streaking on nutrient agar medium plates and were maintained on nutrient agar slants preserved in a refrigerator at 4°C after growth.

4.6 Identification of Phosphate Solubilizing Bacterial Isolates

The isolates were identified on the basis of their morphological, physiological and biochemical characters. The bacterial cultures from the stored nutrient agar slants were sub cultured in the nutrient broth and the broth cultures were subjected to various identification tests. The bacterial cultures were identified using the scheme of Aiso and Simidu (1962). The Bergey's manual (1984) was also referred in the identification procedure.

4.7 Biochemical Tests Employed for the Identification of Phosphate Solubilizing Bacteria

1) Gram's staining technique

A loopful of the broth fresh culture was subjected to Gram staining procedure and observed microscopically under oil immersion objectives.

2) Motility test

The inoculation needle was dipped into the bacterial culture and stabbed into the SIM medium to test for the motility after incubation at 37°C for 24 hours.

3) Penicillin sensitivity test (Kirby-Bauer *et al.*, 1966)

Overnight fresh cultures were streaked on the penicillin sensitive nutrient agar plates. The plates were incubated at 37°C for 24 hours. Then plates were observed for the presence of penicillin resistant isolates.

4) King's media for pigment production test (King's *et al.,* 1954)

From the overnight test bacterial culture, a loopful culture was streaked on the pigment production medium and incubated for 24 hours at 37°C. The results were observed for the pigment production of bacterial isolates.

5) Oxidation-fermentation (OF) test (Hugh and Leifson, 1953)

The oxidation-fermentation basal medium is used for differentiating Gram-negative organisms on the basis of fermentative and oxidative metabolisms of carbohydrates. The oxidative-fermentative basal medium was incorporated with lactose. Culture tube with 2 ml of sterile medium was stabbed with the inoculation needle of dipped bacterial culture and the culture tubes were incubated at 37°C for 24-48 hours. Acid and gas production was observed from the change in the medium colouration and cracking of the medium in the culture tube.

6) Luminescence test

A loopful of test bacterial culture was inoculated into the *Photobacterium* broth and incubated at 37°C for 24 hours. The results were observed for the luminescence of bacterial isolates.

7) Kovac's oxidase test (Kovac's, 1956)

To place the oxidase disc strips into an empty Petri dish and with a sterile platinum wire loop or sterile glass rod smear fresh bacterial culture cells were thoroughly into the oxidase disc. The oxidase test is positive if the transferred cells turn into dark purple colour within 5-10 seconds.

4.8 Growth of Selective Isolates on Selective Media

To confirm the identify and purity of the isolates they were streaked on air dried selective agar medium plates such as Pikovskaya's agar medium and hydroxy appetite agar medium and incubated appropriately. Their growth pattern was observed and recorded.

4.9 Experimental Works

a) Determination of phosphatase enzyme activity in phosphate solubilizing bacteria

i) Preparation of the soup of phosphatase enzyme source

Overnight fresh cultures of phosphate solubilizing bacterial isolates were withdrawn in a 1.5 ml eppendorf and centrifuged at 10,000 rpm for 10 minutes. After centrifugation, the pellet was taken and the soup was discarded. The pellet was resuspended in 1.5 ml of sterile saline. Then this was centrifuged at 10,000 rpm for 10 minutes and the soup was discarded. The pellet was resuspended in 1.5 ml of sterile saline and vortexed well. The enzyme soup was stored at 4°C under refrigeration for the determination of phosphatase enzyme activity.

ii) Assay of phosphatase enzyme on phenolphthalein diphosphate by spectrophotometric method

Phenolphthalein diphosphate (sodium salt), at 0.5% concentration was prepared and filter sterilized. Then, the nutrient broth was prepared and 10ml of this nutrient broth was transferred into test tubes. The nutrient broth test tubes were sterilized by autoclaving. After sterilization, one ml of filter sterilized phenolphthalein diphosphate stock solution was added into each 10 ml of nutrient broth containing test tubes under aseptically and then mixed well. To this, 10 µl of saline suspension of the phosphate solubilizing culture was inoculated aseptically. The test tubes were incubated at 35°C for 24-48 hours. After incubation, 1 ml of 1 N sodium hydroxide (NaOH) solution was added to this incubated test tube cultures. A bright pink red colouration was observed. This colouration indicates the liberation of phenolphthalein due to phosphatase enzymes. Then the optical density of this coloured solution was read at 420 nm in a spectrophotometer (Spectronic 20D).

b) Culture condition optimization for phosphatase enzyme production

i) Effect of pH on phosphatase enzyme production

The phosphate solubilizing bacterial culture that had exhibited maximum phosphatase activity was chosen and subjected to optimization experiments.

Phenolphthalein diphosphate nutrient broth was prepared and pH of the broth medium was adjusted to 5, 6, 7 and 8 using 1 N hydrochloric acid solution (HCl) or 1 N sodium hydroxide solution (NaOH). From each pH group, 10 ml of the broth was transferred into test tubes and sterilized by autoclaving. After sterilization, 100 µl of saline washed cultures were inoculated in all the test tubes aseptically. The tubes were incubated at 37°C for 24 hours and an uninoculated blank was also maintained. A bright pink red colouration was observed. This colouration indicates the liberation of phenolphthalein due to phosphatase enzymes. At six hours interval, phosphatase enzyme activity in its broth culture at different pH was recorded at 420 nm in a spectrophotometer (Spectronic 20D).

ii) Effect of temperature on phosphatase enzyme production

The phosphate solubilizing bacterial isolates that have exhibited high degree of phosphatase activity was chosen and preceded for optimization experiments.

Phenolphthalein diphosphate nutrient broth was prepared and 10 ml of this broth was transferred into test tubes and sterilized by autoclaving. After sterilization, 100 µl of saline washed cultures were inoculated into the test tubes aseptically. Then, one set of culture inoculated broth tubes were incubated at room temperature for 24 hours. Another set of culture inoculated broth tubes were incubated at 35°C for 24 hours in an incubator. Another set of culture inoculated broth tubes were incubated at 45°C for 24 hours in rotator shaker. In each set blank was maintained. A bright pink red colouration was observed. This colouration indicates the liberation of phenolphthalein due to phosphatase enzymes. At six hours interval, broth culture was tested for phosphatase enzyme activity of this coloured solution was read at 420 nm in a spectrophotometer (Spectronic 20D).

c) Evaluation of herbicide resistance in phosphate solubilizer by plate assay method

To check the herbicide resistance profile of the phosphate solubilizing bacterial isolates, plate assay was carried out.

The nutrient agar medium was prepared and sterilized. The concentration herbicide, one shot (2, 4-D and anilofos) was poured into the sterile nutrient agar medium at different concentrations of 10 ppm and 50 ppm aseptically and mixed well. Then, the mixture of the contents were poured (approximately 15-20 ml) into the sterile Petri dishes and was allowed to solidify. After solidification, the saline washed cultures of phosphate solubilizing bacterial isolates were streaked on one shot nutrient agar plates. Then, the plates were incubated at 37°C for 24-48 hours. The control and replicates were maintained. After incubation, the herbicide resistant bacteria isolates were observed and stored for further analysis.

(Aiso and Simudu, 1962)

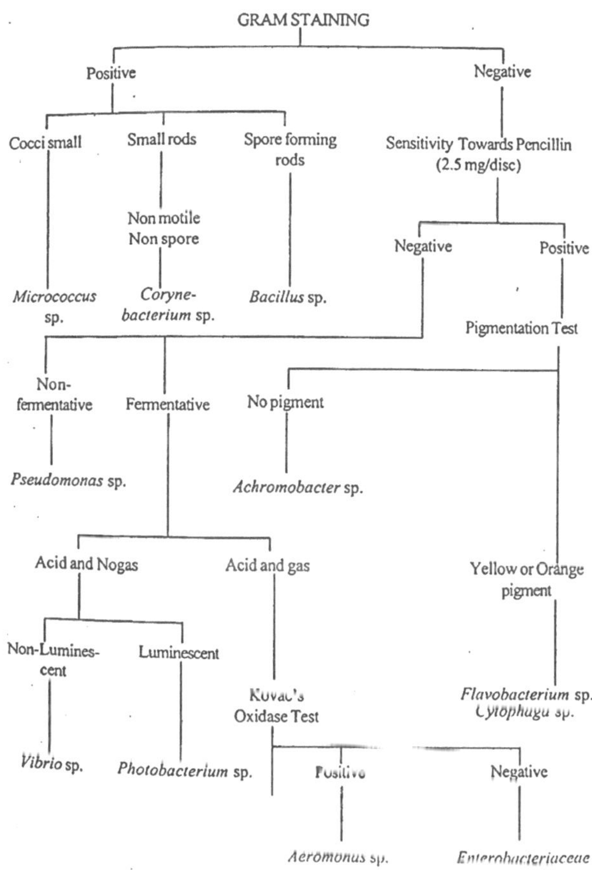

5
RESULTS AND DISCUSSION

Total heterotrophic bacterial population in the rhizosphere soil of *Oryza sativa* was enumerated and the results are presented in table - 1. Throughout the sampling period, total heterotrophic bacterial population was found to be in the range of 10 CFU/gm, similar distribution pattern was recorded in phosphate solubilizing bacteria also (table - 2). All the isolation of phosphate solubilizers was recorded to have comparable recovery of phosphate solubilizers. Comparing the data presented in table - 1 and table - 2, it can be understood that the majority of rhizosphere bacterial isolates were comprised of phosphate solubilizers. Since, the root zone represents high metabolic activity one can expect huge populations that have direct or indirect interaction with plant. Phosphate solubilizers are often found to predominate around rhizosphere region where they could utilize the plant exudates and in turn solubilizer phosphates and make it available for the plants.

Table - 1: Total heterotrophic bacterial population in the rhizosphere soil of *Oryza sativa*

S. No.	Sampling Cycle	Bacterial Load (CFU/gm)	Method of Sampling
01	I	36×10^{-3}	Pour plate technique
02	II	25×10^{-3}	
03	III	32×10^{-3}	

Table - 2: Phosphate solubilizing bacterial load in the rhizosphere soil of *Oryza sativa*

S. No.	Sampling Cycle	Bacterial Load (CFU/gm)			Method of Sampling
		Pikovskaya's Agar Medium	Hydroxy Apatite Agar Medium	Soil Extract Agar Medium	
01	I	50×10^{-2}	30×10^{-2}	20×10^{-2}	Spread plate technique
02	II	30×10^{-2}	20×10^{-2}	20×10^{-2}	
03	III	30×10^{-2}	30×10^{-2}	10×10^{-2}	

On the basis of phosphate solubilization, twenty five bacterial isolates were characterized up to generic level and the data was presented in table - 3. From this, it can be noted that the major components are *Micrococcus* sp., *Bacillus* sp. and *Pseudomonas* sp. Among these, both *Micrococcus* sp. and *Bacillus* sp. were equally distributed (48%) and the only other isolate was *Pseudomonas* sp. (4%) presented in figure - 1.

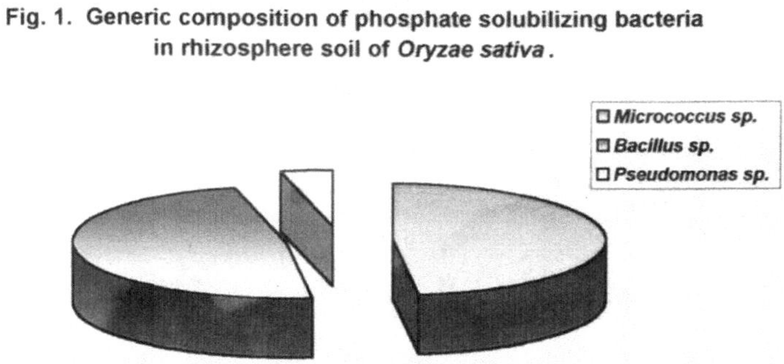

Figure 1: Generic composition of phosphate solubilizing bacteria in rhizosphere soil of *Oryza sativa*

S. No.	Tests Performed	Phosphate Solubilizing Bacterial Isolates																								
		1	2	3	4	5	6	7	8	9	10	11	12	13	14	15	16	17	18	19	20	21	22	23	24	25
01	Morphology	Small cocci	Rod shaped	Rod shaped	Rod shaped	Small cocci	Small cocci	Small cocci	Small cocci	Small cocci	Small cocci	Rod shaped	*Rod shaped	Rod shaped	Small cocci	Rod shaped	Rod shaped	Rod shaped	Rod shaped	Small cocci	Rod shaped	Rod shaped	Small cocci	Small cocci	Small cocci	Rod shaped
02	Gram staining	+	+	+	+	+	+	+	+	+	+	+	-	+	+	+	+	+	+	+	+	+	+	+	+	+
03	Motility	-	+	+	+	-	-	-	-	-	-	+	+	+	-	+	+	+	+	-	+	+	-	-	-	+
04	Penicillin sensitivity test	-	-	-	-	-	-	-	-	-	-	-	+	-	-	-	-	-	-	-	-	-	-	-	-	-
05	Pigment production test	-	-	-	-	-	-	-	-	-	-	-	-	-	-	-	-	-	-	-	-	-	-	-	-	-
06	Oxidation-fermentation test	-	-	-	-	-	-	-	-	-	-	-	-	-	-	-	-	-	-	-	-	-	-	-	-	-
07	Luminescence test	-	-	-	-	-	-	-	-	-	-	-	-	-	-	-	-	-	-	-	-	-	-	-	-	-
08	Kovac's oxidase test	-	-	-	-	-	-	-	-	-	-	-	-	-	-	-	-	-	-	-	-	-	-	-	-	-

Table – 3: Biochemical characterization of phosphate solubilizing bacterial isolates

*Legend: Small cocci indicates Micrococcus sp., rod shaped indicates Bacillus sp., and *rod shaped indicates Pseudomonas sp.*

Growth of this organisms in media supplemented with inorganic phosphate indicates their ability to secrete organic acids that facilitate phosphate solubilization. Most phosphate solubilizing organisms have capabilities of elaborating enzyme namely phosphatase that plays a vital role in the catalysis of organic phosphates.

To understand this phenomenon, five isolates were selected at random (*Bacillus* sp. PS 11, PS 16, PS 18, PS 25 and *Micrococcus* sp. PS 23) and examined for phosphatase activity with the help of an appropriate indicator. Photometric analysis of this experiment was presented in figure - 2. From this, it can be understood that *Bacillus* sp. PS 11, PS 18 and PS 25 were slightly more efficient in phosphatase production compared to the other two.

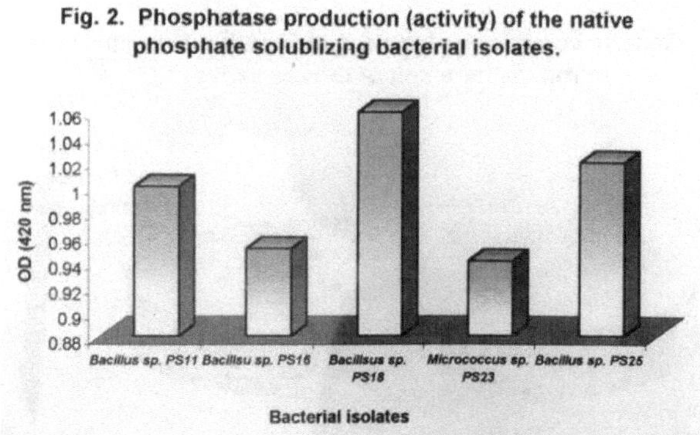

Figure 2: Photometric analysis of phosphatase enzyme production (activity) of the native phosphate solubilizing bacterial isolate

Further all the five strains were subjected to culture optimization experiments and the data are presented in figure 3-12. In the case of phosphate solubilizer (PS) 11, optimum pH for enzyme production was recorded to be 7.0 (figure - 3) and the temperature at 45°C (figure - 4).

Similar observations can be noted for *Bacillus* sp. PS 16, PS 18 and *Micrococcus* sp. PS 23 (figure - 5-10)

In the case of *Bacillus* sp. PS 25, the optimum pH for phosphatase production was recorded to be pH 8.0 and the temperature of 35°C.

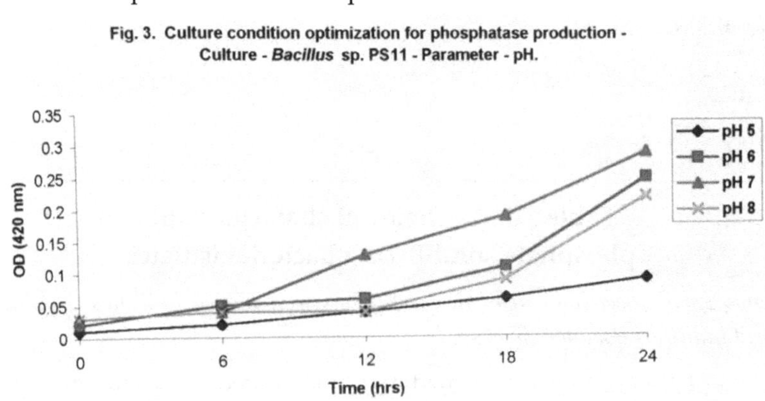

Figure 3: Culture condition optimization for phosphatase production of *Bacillus* sp. PS 11 in the parameter of pH

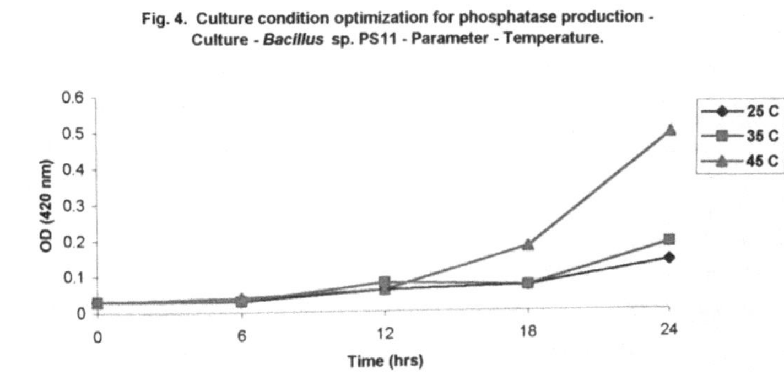

Figure 4: Culture condition optimization for phosphatase production of *Bacillus* sp. PS 11 in the parameter of temperature

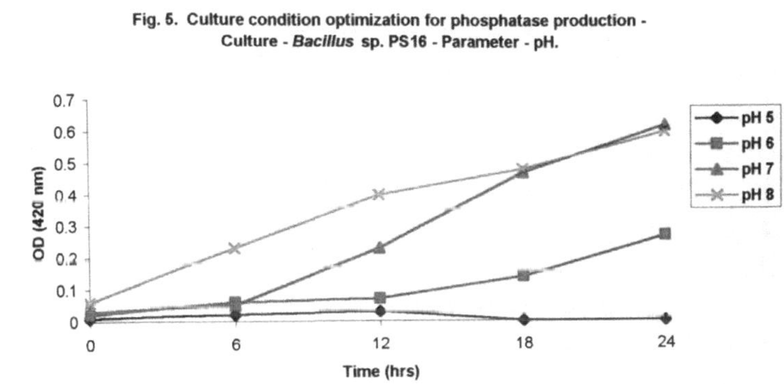

Figure 5: Culture condition optimization for phosphatase production of *Bacillus* sp. PS 16 in the parameter of pH

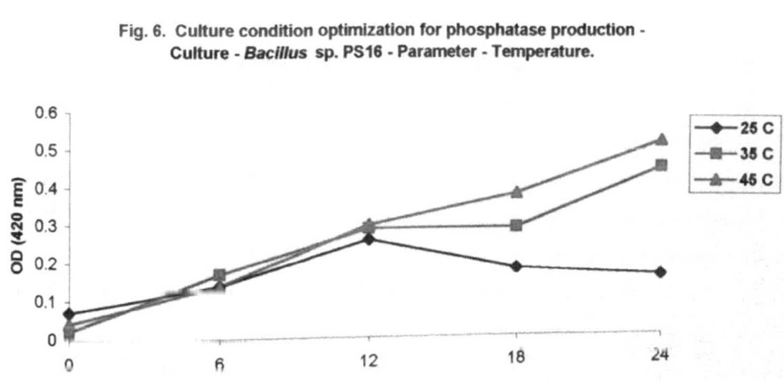

Figure 6: Culture condition optimization for phosphatase production of *Bacillus* sp. PS 16 in the parameter of temperature

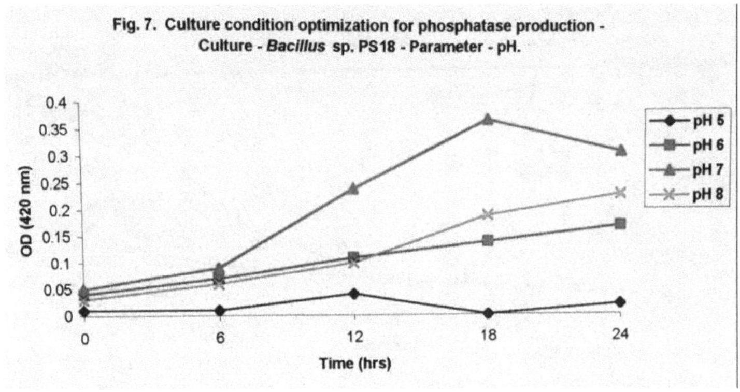

Figure 7: Culture condition optimization for phosphatase production of *Bacillus* sp. PS 18 in the parameter of pH

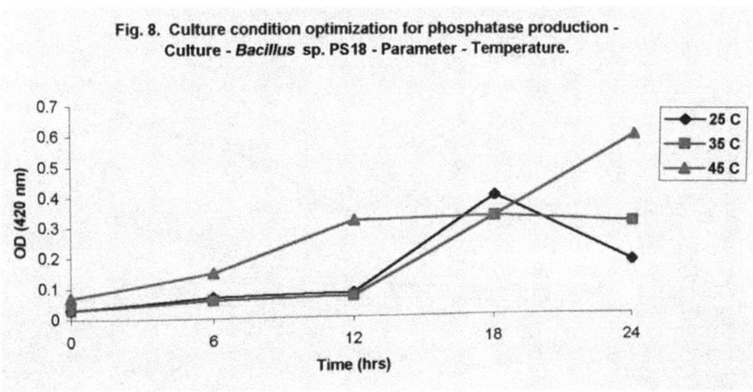

Figure 8: Culture condition optimization for phosphatase production of *Bacillus* sp. PS 18 in the parameter of temperature

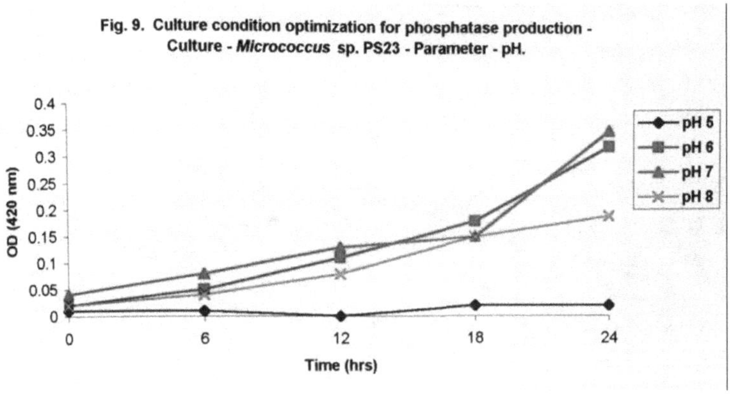

Figure 9: Culture condition optimization for phosphatase production of *Micrococcus* sp. PS 23 in the parameter of pH

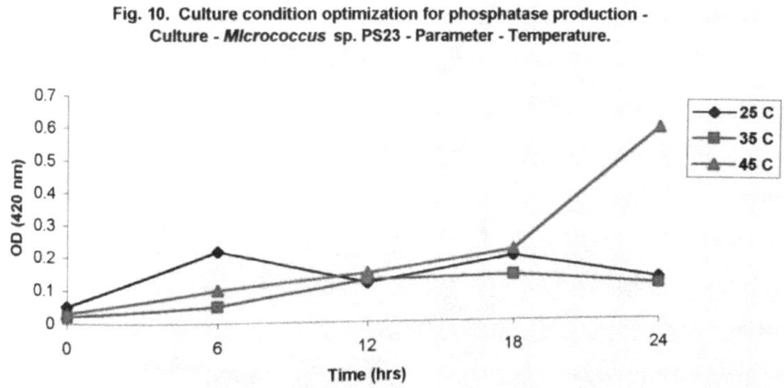

Figure 10: Culture condition optimization for phosphatase production of *Micrococcus* sp. PS 23 in the parameter of temperature

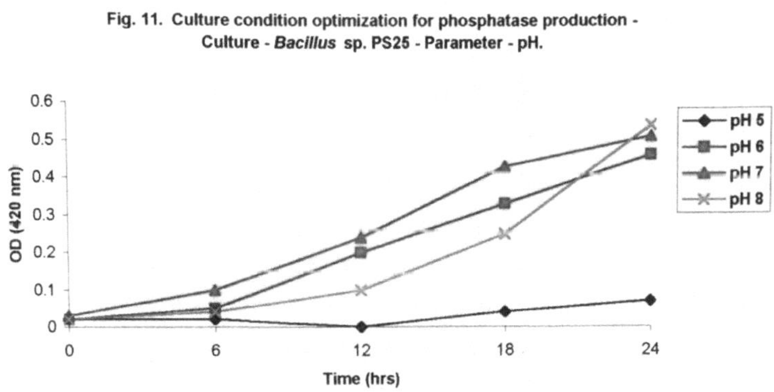

Figure 11: Culture condition optimization for phosphatase production of *Bacillus* sp. PS 25 in the parameter of pH

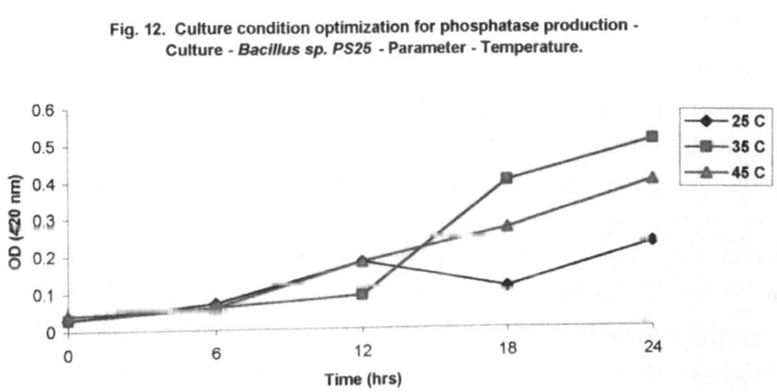

Figure 12: Culture condition optimization for phosphatase production of *Bacillus* sp. PS 25 in the parameter of temperature

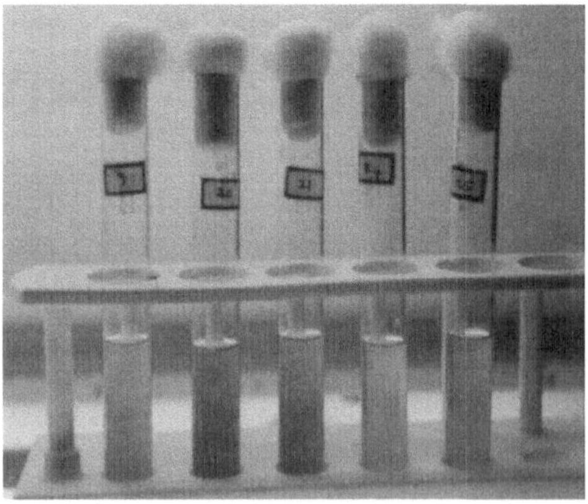

Figure 13: Phosphatase enzyme assay (before adding 1 N NaOH solution)

Legend: Light pink colour indicates the presence of phosphatase enzyme

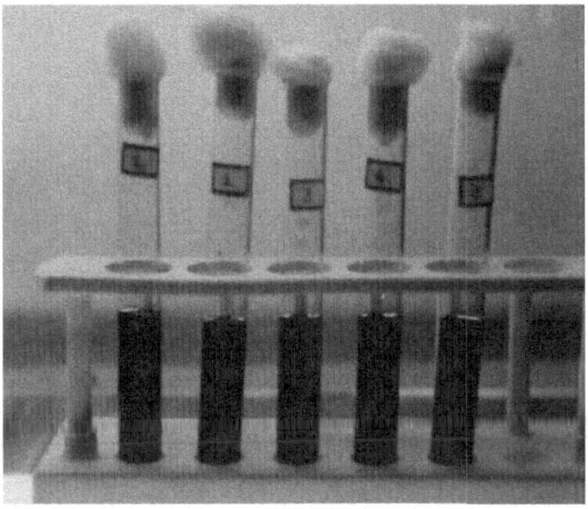

Figure 14: Phosphatase enzyme assay (after adding 1 N NaOH solution)

Legend: Bright pink colour indicates the presence of phosphatase enzyme

Phosphatase production is an important property that enables the bacteria to flourish especially in the rhizosphere region. As plants are continuously amended with organic fertilizers, these organisms would generate phosphate by more than one mechanism. Furthermore, phosphatase producers would be of special value especially when organophosphate pesticides through phosphatase enzyme thereby ensure the faster disappearance in soil. Even though these organisms are expected to resist organophosphonates, resistance to other pesticides would be an added advantage as these organisms can be used both as biofertilizers and bioremedial agents. In this background, one isolate from each genera was exposed to a commonly used herbicide, 'one shot' (2, 4-D and anilofos) at two different concentrations (10 ppm and 50 ppm) and checked for their tolerance level, while *Micrococcus* sp. and

Bacillus sp. were unable to tolerate this herbicide even at low concentration (10 ppm). *Pseudomonas* sp. was able to tolerate this farm chemical at lower concentration (table - 4).

Table - 4: Herbicide (one shot) resistance pattern of phosphate solubilizing bacterial isolates

S. No.	Culture Tested	Concentration of Herbicide	
		10 ppm	50 ppm
01	*Micrococcus sp.*	-	-
02	*Bacillus sp.*	-	-
03	*Pseudomonas sp.*	+	-

Legend: sign indicates the appearance of bacterial growth and - sign indicates the absence of bacterial growth.

6
CONCLUSION

This study indicates, the abundance of phosphate solubilizing bacteria in the rhizosphere region of *Oryza sativa*. Furthermore, their phosphatase activity would prove beneficial especially in inactivating residual organophosphate pesticides that would prevent the problem of pesticide pollution due to this chemical. Further research on the nature and production of phosphatase in these strains would be useful in their commercial applications.

7
BIBLIOGRAPHY

1. Goldstein AH. Bacterial solubilization of mineral phosphates: Historical perspective and future prospects. Am J Alt Agric, March Spring 1986; 1(2): 51-57.https://doi.org/10.1017/S0889189300000886.
2. Gyaneshwar P, Naresh Kumar G, Parekh LJ. Effect of buffering on the phosphate solubilizing ability of microorganisms. World J Microbiol Biotech, October 1998; 14: 669-673. https://doi.org/10.1023/A:1008852718733.
3. Kennedy AC. Rhizosphere. In: Principles and applications of soil microbiology, Sylvia DM, Fuhrmann JJ, Hartel PG, Zuberer DA (eds.). 2nd ed., Pearson Prentice Hall, New Jersey; 2005, pp. 242-262.
4. Hiltner L. Uber neuere erfahrungen und probleme auf dem gebiet der bodenbakterio logie und unter beson derer berucksichtigung der grundungung and brache. Arbeiten der Deutschen Landwirtschaftlichen-Gesellschaft, 1904; 98: 59-78.
5. Vessey JK. Plant growth promoting rhizobacteria as biofertilizers. Plant and Soil, 2003; 255: 571-586. http://dx.doi.org/10.1023/A:1026037216893.
6. Antoun H, Kloepper JW. Plant growth-promoting rhizobacteria. In: Brenner S, and Miller JH (eds.). Encyclopedia of Genetics, Academic Press, New York; 2001, pp. 1477-1480.
7. Goldstein AH, Rogers RD, Mead G. Mining by microbe. Nature Biotechnology, November 1993; 11: 1250-1254.
8. Fasim F, Ahmed N, Parsons R, Gadd GM. Solubilization of zinc salts by a bacterium isolated from the air environment of a tannery. FEMS Microbiology Letters, July 2002; 213(1): 1-6. https://doi.org/10.1111/J.1574-6968.2002.tb11277.x.
9. Khan MS, Zaidi A, Wani PA. Role of phosphate solubilizing microorganisms in sustainable agriculture - a review. Agronomy for Sustainable Development, March 2007; 27: 29-43. https://doi,org/10.1051/agro:2006011.
10. Glick BR. The enhancement of plant growth by free-living bacteria. Canadian Journal of Microbiology, February 1995; 41(2): 109-117. https://doi.org/10.1139/m95-015.

11. Iguala JM, Valverde A, Cervantes E, Velazquez E. Phosphate solubilizing bacteria as inoculants for agriculture: use of updated molecular techniques in their study. Agronomie, September 2001; 21(6): 561-568. https://doi.org/10.1051/agro:2001145.

12. Wu SC, Cao ZH, Li ZG, Cheung KC, Wong MH. Effects of biofertilizer containing N-fixer, P and K solubilizers and AM fungi on maize growth: a greenhouse trial. Geoderma, March 2005; 125(1/2): 155-166. https://doi.org/10.1016/j.geoderma.2004.07.003.

13. Donahue RL, Miller RW, Shickluna JC. Soils: An introduction to soils and plant growth. New Delhi - 110 001, Prentice Hall of India Private Limited: 1990, pp. 222-4.

14. Illmer P, Schinner F. Solubilization of inorganic phosphates by microorganisms isolated from forest soils. Soil Biology and Biochemistry, April 1992; 24: 389-395. https://doi.org/10.1016/0038-0717(92)90199-8.

15. Tewari SK, Das B, Mehrotra S. Cultivation of medicinal plants-tool for rural development. J Rural Tech, 2004; 3: 147-50.

16. Neelima Garg M, Mishra M, Garg KL. Microbial solubilization of insoluble phosphates. In: Mukerji KG, Singh VP and Garg KL (eds.). Frontiers in Applied Microbiology, New Delhi; 1989, Rastogi and Company.

17. Skujins J. Enzymes in soil. In: McLaren D and Peterson GH (eds.). Soil Biochemistry, Vol. 1, Marcel Dekker, New York; 1967, pp. 371-414.

18. Sen A, Paul NB. Solubilization of phosphates by some common soil bacteria. Current Science, 1957; 26: 222-223.

19. Swaby RJ, Sperber JI. Phosphate dissolving microorganisms in the rhizosphere of legumes. In: nutrition of the legumes. New York; Academic Press: 1959, pp. 289-294.

20. Sperber JI. Solution of mineral phosphates by soil bacteria. Nature, November 1957; 180(4593): 994-995. https://doi.org/10.1038/180994a0.

21. Artidave A, Patel HH. Inorganic phosphate solubilizing *Pseudomonas*. Indian Journal of Microbiology, 1999; 39: 161-164.

22. Miligi L, Costantini AS, Veraldi A, Benvenuti A, Vineis P. Cancer and pesticides: an overview and some results of the Italian multicenter case-control study on hemato lymphopoietic malignancies. Annals of the New York Academy of Sciences, October 2006; 1076: 366-377. https://doi.org/10.1196/annals.1371.036.

23. Vyas SC. Soil microorganisms and their activities. Nontarget effects of agricultural fungicides, CRC Press, FL, Boca Raton, 1988, pp. 258-275.

24. Edwards CA, Bater JE. An evaluation of laboratory and field studies for assessment of the environmental effects of pesticides. Proceedings of the Brighton Crop Production Conference on Pests and Diseases, 1990; 963-968.

25. Khan M, Scullion J. Effect of soil on microbial responses to metal contamination. Environmental Pollution, October 2000; 110(1): 115-125. https://doi.org/10.1016/S0269-7491(99)00288-2.

26. Srinivas T, Sridevi M, Mallaiah KV. Effect of pesticides on *Rhizobium* and nodulation of green gram *Vigna radita* (L.) Wilczek. The IUP Journal of Life Sciences, 2008; 2: 36-44.

27. Pereira JL, Picanco MC, Silva AA, Santos EA, Tome HVV, Olarte JB. Effects of glyphosate and endosulfan on soil microorganisms in soybean crop. Planta Daninha, October 2008; 26(4): 825-30. http://doi:10.1590/S0100-83582008000400014.

28. Ahemad M, Khan MS, Zaidi A, Wani PA. Remediation of herbicides contaminated soil using microbes. In: Khan MS, Zaidi A and Musarrat J (eds.). Microbes in Sustainable Agriculture, New York; Nova Publishers: 2009.

29. Barriuso J, Marin S, Mellado RP. Effect of the herbicide glyphosate on glyphosate- tolerant maize rhizobacterial communities: a comparison with pre-emergency applied herbicide consisting of a combination of acetochlor and terbuthylazine. Environ Microbiol, 2010; 12(4): 1021-30. http://doi:10.1111/j.1462-2920.2009.02146.x.

30. Ahemad M, Khan MS. Effect of insecticide-tolerant and plant growth promoting Mesorhizobium on the performance of chickpea grown in insecticide stressed alluvial soils. J Crop Sci Biotech, December 2009; 12(4): 213-222. http://doi:10.1007/s12892-009-0130-8.

31. Srinivas T, Sridevi M, Mallaiah KV. Effect of pesticides on *Rhizobium* and nodulation of green gram *Vigna radita* (L.) Wilczek. The IUP Journal of Life Sciences, 2008; 2: 36-44.

32. Ahemad M, Khan MS. Phosphate solubilizing and plant growth promoting *Pseudomonas aeruginosa* PS1 improves green gram performance in quizalafop-p-ethyl and clodinafop amended soil. Arch Environ Contam Toxicol, February 2010; 58(2): 361-372. http://doi:10.1007/s00244-009-9382-z.

33. Bellinaso ML, Greer CW, Peralba MC, Henriques JA, Gaylarde CC. Biodegradation of the herbicide trifluralin by bacteria isolated form soil. FEMS Microbiol Ecol, March 2003; 1, 43(2): 191-194. http://doi:10.1111/j.1574-6941.2003.tb01058.x.

34. Wani PA, Zaidi A, Khan AA, Khan MS. Effect of phorate on phosphate solubilization and indole acetic acid releasing potentials of rhizospheric microorganisms. Ann Plant Protec Sci, January 2005; 13: 139-144.

35. Ahemad M, Khan MS. Effect of insecticide-tolerant and plant growth promoting Mesorhizobium on the performance of chickpea grown in insecticide stressed alluvial soils. J Crop Sci Biotech, December 2009; 12(4): 213-222. http://doi:10.1007/s12892-009-0130-8.

36. Oves M, Zaidi A, Khan MS, Ahemad M. Variation in plant growth promoting activities of phosphate solubilizing microbes and factors affecting their colonization and solubilizing efficiency in different agro-ecosystems. In: Khan MS and Zaidi A (eds.). Phosphate Solubilizing Microbes for Crop Improvement, New York; Nova Science Publishers Inc: July 2009, pp. 247-263.

37. Aiso K, Simidu V. Occurrence and distribution of heterotrophic bacteria in seawater from the komagawa bay. Bull Jap Soc Sci Fish, 1962; 28: 1133-1141.

38. Darmwall NS, Singh RB, Rai R. Isolation of phosphate solubilizers from different sources. Curr Sci, 1989; 58: 570-571.

39. Gaur AC, Sachar S. Effect of rock phosphate on phosphate solubilization by *Aspergillus awamori*. Current Science, 1980; 49(1980): 553-554.

40. Ahamed N, Jha KK. Solubilization of rock phosphate by micro-organisms isolated from Bihar soils. J Gen Appl Microbiol, 1968; 14: 89-95.

41. Bhurat MC, Sen A. Phosphate solubilizing bacteria in phyllosphere of winer crops. Indian Journal of Microbiology, 1968; 8(21): 255-256.

42. Thakkar J, Narsian V, Patel HH. Inorganic phosphate solubilization by certain soil bacteria. Indian J Exp Biol, 1993; 31: 743-746.

43. Katznelson H, Bose B. Metabolic activity and phosphate-dissolving capability of bacterial isolates from wheat roots, rhizosphere, and non-rhizosphere soil. Canadian Journal of Microbiology, 1959; 5(1959): 79-85.

44. Goswami KP, Sen A. Solubilization of calcium phosphate by three strains of phosphobacteria. Indian J Agri Sci, 1962; 32: 96–101.

45. Kucey RMN. Phosphate-solubilizing bacteria and fungi in various cultivated and virgin Alberta soils. Canadian Journal of Soil Science, November 1983; 63(4): 671-678. https://doi.org/10.4141/cjss83-068.

46. Cunningham JE, Kuiack C. Production of citric and oxalic acids and solubilization of calcium phosphate by *Penicillium bilaii.* Applied and Environmental Microbiology, 1992; 58(5): 1451-1458. https://doi.org/10.1128/aem.58.5.1451-1458.1992.

47. Thakker J, Narsian V, Patel HH. Inorganic phosphate solubilization by certain soil bacteria. Indian J Exp Biol, 1993; 31: 743-746.

48. Illmer PA, Schinner F. Solubilization of inorganic calcium phosphates solubilization mechanisms. Soil Biology and Biochemistry, 1995; 27: 257-263. https://doi.org/10.1016/0038-0717(94)00190-c.

49. Sudhansu S Pal. Selection of acid tolerant strains of phosphate-solubilizing bacteria in soils of Uttar Pradesh Himalaya. Indian Journal of Agricultural Science, August 1999; 69(8): 578-582.

50. Artidave A, Patel HH. Inorganic phosphate solubilizing *Pseudomonas*. Indian J Microbiology, 1999; 39: 161-164.

51. Alvaro Peix, Raul Rivas, Pedro F. Mateos, Eustoquio Martinez-Molina, Claudino Rodriguez-Barrueco and Encarna Velazquez. *Pseudomonas rhizosphaerae* sp. Nov., a novel species that actively solubilizes phosphate *in vitro*. International Journal of Systematic and Evolutionary Microbiology, June 2003; 53: 2067-2072. http://doi.org/10.1099/ijs.0.02703-0.

52. Chung H, Park M, Madhaiyan M, Seshadri S, Song J, Cho H, Sa T. Isolation and characterization of phosphate solubilizing bacteria from the rhizosphere of crop plants of Korea. Soil Biology and Biochemistry, 2005; 37: 1970-1974. https://doi.org/10.1016/j.soilbio.2005.02.025.

53. Henri Fankem, Dieudonne Nwaga, Annette Deubel, Lamine Dieng, Wolfgang Merbach, Francois Xavier Etoa. Occurrence and functioning of phosphate solubilizing microorganisms from oil palm tree (Elaeis guineensis) rhizosphere in cameroon. African Journal of Biotechnology, December 2006; 5(24): 2450-2460.

54. Amal Chandra Das, Anjan Debnath. Effect of systemic herbicides on N2-fixing and phosphate solubilizing microorganisms in relation to availability of nitrogen and phosphorus in paddy soils of West Bengal. Chemosphere, November 2006; 65(6): 1082-1086. https://doi.org/10.1016/j.chemosphere.2006.02.063.

55. Narula N, Deubel A, Gransee A, Merbach W. Phosphate solubilizing microorganisms: mechanism of solublilization as influenced by pH, type of organic acid and sugars. Asian Journal of Microbiology, Biotechnology and Environmental Sciences, 2007; 9(2): 397-404.

56. Sadaf Shahab, Nuzhat Ahmed. Effect of various parameters on the efficiency of zinc phosphate solubilization by indigenous bacterial isolates. African Journal of Biotechnology, 2008; 7(10): 1543-1549.

57. Joseph S, Jisha MS. Buffering reduces phosphate solubilizing ability of selected strains of bacteria. World J Agric Sci, 2009; 5(1): 135-137.

58. Arvind Kumar, Poonam Bhargava, Lal Chand Rai. Isolation and molecular characterization of phosphate solubilizing *Enterobacter* and *Exiguobacterium* species from paddy fields of Eastern Uttar Pradesh, India. African Journal of Microbiology Research, 4 May 2010; 4(9): 820-829.

59. Tabatabai MA, Brenner JM. Use of p-nitrophenol phosphate for the assay of soil phosphatase activity. Soil Biology Biochemistry, 1969; 1: 301-307. http://dx.doi.org/10.1016/0038-0717(69)90012-1.

60. Ayyakkannu K, Chandramohan D. Occurrence and distribution of phosphate solubilizing bacteria and phosphatase in marine sediments at Porto Novo. Marine Biology, November 1971; 11: 201-205.

61. Richard G, Tigerstrom Von. Production of two phosphatases by *Lysobacter enzymogenes* and purification and characterization of the extracellular enzyme. Applied and Environmental Microbiology, 1984; 47: 693-698. https://doi.org/10.1128/AEM.47.4.693-698.1984.

62. De Prada P, Loveland-Curtze J, Brenchley JE. Production of two extracellular alkaline phosphatases by a psychrophilic *Arthrobacter* strain. Applied and Environmental Microbiology, 1996; 62(10): 3732-3738. https://doi.org/10.1128/aem.62.10.3732-3738.1996.

63. Rodríguez H, Fraga R. Phosphate solubilizing bacteria and their role in plant growth promotion. Biotechnology Advances, October 1999; 17(4-5): 319-339. https://doi.org/10.1016/s0734-9750(99)00014-2.

64. Zhong ChuanQing, Huang Wei Yi. Comparison in phosphate solubilizing effects between different phosphate solubilizing microbes and variation of activities of their phosphatases. Acta Pedologica Sinica, 2005; 42(2): 286-294. https://doi.org/10.11766/trxb200403150217.

65. Ponmurugan P, Gopi C. Distribution pattern and screening of phosphate solubilizing bacteria isolated from different food and forage crops. Journal of Agronomy, 2006; 5(4): 600-604. https://doi.org/10.3923/ja.2006.600.604.

66. Sundara Rao WVB, Sinha MK. Phosphate dissolving microorganisms in the soil and rhizosphere. Indian J Agric Sci, April 1963; 33: 272-278.

67. Ayyakkannu K, Chandramohan D. On the occurrence and distribution of phosphobacteria in the marine environment at Porto Novo. Curr Sci, September 1970; 39(17): 398-399. http://doi.org/www.jstor.org/stable/24074482.

68. Katznelson H, Bose B. Metabolic activity and phosphate-dissolving capability of bacterial isolates from wheat roots, rhizosphere, and non-rhizosphere soil. Can J Microbiol, February 1959; 5(1): 79-85. http://doi.org/10.1139/m59-010.

69. Bergey's Manual of Systematic Bacteriology. Baltimore MD, Williiams and Wilkins, 1984-1989.

70. Bauer AW, Kirby WM, Sherris JC, Turck M April. Antibiotic susceptibility testing by a standardized single disk method. Am J Clin Pathol, April 1996; 45(4): 493-496. http://doi.org/10.1093/AJCP/45.4-TS.493.

71. King ED, Ward MK, Raney DE. Two simple media for the demonstration of pyo-cyanin and fluorescin. J Lab and Clin Med, August 1954; 44(2): 301-307. http://doi.org/10.5555/URI:PII:002221435490222X.

72. Hugh R, Leifson E. The taxonomic significance of fermentative versus oxidative metabolism of carbohydrates by various Gram-negative bacteria. J Bacteriol, July 1953; 66(1): 24-26. http://doi.org/10.1128/jb.66.1.24-26.1953.

73. Kovacs N. Identification of *Pseudomonas pyocyanea* by the oxidase reaction. Nature (London), September 1956; 178(4535): 703. http://doi.org/10.1038/178703a0.

APPENDIX

Gram Stains - Kit

Intended Use

Grams stain kit is used for differentiation of bacteria on the basis of their Gram nature.

Composition

Ingredients

Gram's Crystal Violet (Solution A)

Crystal violet	- 2.0 g
Ethyl alcohol 95%	- 20.0 ml

Gram's Crystal Violet (Solution B)

Ammonium oxalate, monohydrate	- 0.80 g
Distilled water	- 80.0 ml

Solution A and B mixed. It is stored for 24 hours before use. The resulting stain is stable.

Gram's Decolourizer

Ethyl alcohol, 95%	- 50.0 ml
Acetone	- 50.0 ml

Gram's Iodine

Iodine	- 1.0 g
Potassium iodide	- 2.0 g
Distilled water	- 300.0 ml

Safranin, 0.5% w/v

Safranin O	**- 0.50 g**
95% ethyl alcohol	- 100.0 ml

Principle and Interpretation

The Gram stain is a differential staining technique most widely applied in all microbiology disciplines laboratories. It is one of the most important criteria in any identification scheme for all types of bacterial isolates. Different mechanisms have been proposed to explain the Gram reaction. There are many physiological differences between Gram-positive and Gram-negative cell walls. Ever since Christian Gram has discovered Gram staining, this process has been extensively investigated and redefined. In practice, a thin smear of bacterial cells is stained with crystal violet, and then treated with an iodine containing mordant to increase the binding of primary stain. A decolourizing solution of alcohol or acetone is used to remove the crystal violet from cells which bind it weakly and then the counter stain (like safranin) is used to provide a colour contrast in those cells that are decolourized. Gram-positive bacteria have a thick mesh-like cell wall made of peptidoglycan (50-90% of cell envelope), and as a result are stained purple by crystal violet, whereas Gram-negative bacteria have a thinner layer (10% of cell envelope), so do not retain the purple stain and are counter-stained pink by safranin. In a properly stained smear by Gram staining procedure, the Gram-positive bacteria appear blue to purple and Gram-negative cells appear pink to red.

Hydroxy Apetite Medium

Components	Gms / Litre
Yeast extract	0.2
Ammonium sulfate	0.5
Magnesium sulfate	0.1
Potassium chloride	0.2
Dextrose	10.0
Soil extract	200 ml
Distilled water	800 ml
Trace elements solution	1.0 ml
Agar	20.000
pH (at 25°C)	7.2 ± 0.1

After sterilization add 60 ml of 10% calcium chloride and 40 ml of 10% potassium dihydrogen phosphate. Adjust the pH with 0.1 N sodium hydroxide solutions.

Preparation of Trace Elements Solution

Components	Grams / Litre
Boric acid	0.5
Ammonium molybdate	0.5
Sodium bromide	0.02
Zinc sulfate	0.02
Ammonium sulfate	0.30
Soil extract	200 ml

King's Medium B Base

Intended Use

Kings medium B base is recommended for non-selective isolation, cultivation and pigment production of *Pseudomonas* species.

Composition

Ingredients	Gms / Litre
Proteose peptone	20.000
Dipotassium hydrogen phosphate	1.500
Magnesium sulphate heptahydrate	1.500
Agar	20.000
Final pH (at 25°C)	7.2 ± 0.2

Principle and Interpretation

Pseudomonas aeruginosa is known to produce two types of pigments, pyocyanin and fluorescein which is a characteristic property and aidsin isolation of *Pseudomonas* from clinical material. An additional pigment called as pyorubin was reported by King. Pyocyanin is green while fluorescein is fluorescent yellow and pyorubin is reddish brown. Some strains produce all these pigments while the others produce one or two pigments.

Pseudomonas aeruginosa can be identified on Hugh-Leifson medium. Kings medium B base is particularly suited for fluorescein.

Kings medium B base is based on the formulation of King *et al*. This medium can be used as a general medium for the non-selective isolation and pigment production of *Pseudomonas* species from foods, cosmetic samples etc. *Agrobacterium* have been traditionally identified as Gram-negative bacteria that do not produce fluorescent pigment on Kings B medium and do produce tumors (or hairy roots) when inoculated on to test plants.

These media contain proteose peptone, which provides carbonaceous and nitrogenous compounds for the growth of bacteria. Glycerol serves as a source of energy and also enhances pigment production. Magnesium sulphate also enhances pigment production. Pigments and / or their derivatives produced by *Pseudomonas* species play a role as siderophores in their own uptake systems of bacteria, and hence, their production is markedly enhanced under conditions of iron deficiency. The production of pigments especially non-fluorescent blue pigment, pyocyanin is readily demonstrated by culturing on Kings medium B, which contains no added iron. The addition of dipotassium phosphate increases the phosphorus content of the medium there by enhancing production of fluorescent pigment. For inoculation, use the organisms freshly cultured in Kings medium A, incubate overnight at 37°C and then at room temperature for six days. With Kings medium B, incubate at 37°C for six days.

Gordon-McLeod Reagent (Oxidase reagent)

Intended Use

Cordon-McLeod reagent (Oxidase reagent) is used to carry out the oxidase test, to determine the presence of oxidase enzymes.

Composition

Ingredients

N, N-Dimethyl p-phenylene diamine hydrochloride	-0.15 gm
Distilled water	-10.0 ml

Principle and Interpretation

The oxidase test is based on bacterial production of an intracellular oxidase enzyme. This oxidase reaction is due to a cytochrome oxidase system that activates oxidation of reduced cytochrome by molecular oxygen, which in turnacts as an electron acceptor in the terminal stage of the electron transfers system. All *Pseudomonas* and *Neisseria* sp., produce an oxidase enzyme which, in the presence of atmospheric oxygen, cytochrome c, and an oxidase reagent, oxidize the reagent to a colored compound, indophenol.

Mueller-Hinton Agar

Intended Use

Mueller-Hinton agar is used for determination of susceptibility of microorganisms to antimicrobial agents.

Composition

Ingredients	Gms / Litre
Beef, infusion from	300.000
Casein acid hydrolysate	17.500
Starch	1.500
Agar	20.000
Final pH (at 25°C)	7.3 ± 0.1

Principle and Interpretation

The Mueller-Hinton formulation was originally developed as a simple, transparent agar medium for the cultivation of pathogenic *Neisseria* species. Other media were subsequently developed that replaced the use of Mueller-Hinton agar for the cultivation of pathogenic *Neisseria* species, but it became widely used in the determination of sulfonamide resistance of Gonococci and other organisms. Mueller-Hinton agar is now used as a test medium for antimicrobial susceptibility testing. Mueller-Hinton agar is recommended for the diffusion of antimicrobial agents impregnated on paper disc through an agar gel as described in CLSI approved standard. Mueller-Hinton agar has been selected by the CLSI for several reasons:

i. It demonstrates good batch-to-batch reproducibility for susceptible testing.

ii. It is low in sulfonamide, tri methoprim and tetracycline inhibitors.

iii. It supports the growth of most non-fastidious bacterial pathogens and

iv. Many data and much experience regarding its performance have been recorded.

Kirby-Bauer *et al.,* recommended this medium for performing antibiotic susceptibility tests using a single disc of high concentration. WHO committee on standardization of susceptibility testing has accepted Mueller-Hinton agar for determining the susceptibility of microorganisms because of its reproducibility. Mueller-Hinton agar with 5% sheep blood and Mueller-Hinton agar with hemoglobin have been recommended for antimicrobial susceptibility testing of *Streptococcus pneumoniae* and *Haemophilus influenzae.*

Beef infusion and casein acid hydrolysate provide nitrogenous compounds, carbon, sulphur and other essential nutrients. Starch acts as a protective colloid against toxic substances present in the medium. Starch hydrolysis yields dextrose, which serves as a source of energy. These ingredients are selected for low thymine and thymidine content as determined by MIC values for *Enterococcus faecalis* with sulfamethoxazole-trimethoprim (SXT). Calcium and magnesium ion concentrations are adjusted to provide the amounts recommended by CLSI to give the correct MIC values with amino glycosides and *Pseudomonas aeruginosa.*

The Kirby-Bauer procedure is based on agar diffusion of antimicrobial substances impregnated on paper discs. This method employs disc with a single concentration of antimicrobial agent and the zone diameters observed are correlated with minimum inhibitory concentration (MIC) values. A standardized suspension of the organism is swabbed over the entire surface of the medium. Paper discs impregnated with specific amounts of antimicrobial agents are then placed on the surface of the medium, incubated and zones of inhibition around each disc are measured. The susceptibility is determined by comparing with CLSI standards. The various factors, which influence disc diffusion susceptibility tests, are agar depth, disc potency, inoculums concentration, pH of the medium and beta-lactamase production by test organisms.

Mueller-Hinton agar is not appropriate for assay by disc diffusion method with slow growing organisms, anaerobes and capnophiles. With slow growing organisms, increased incubation may cause deterioration of diffusing antibiotic and produce unprecise readings.

Nutrient Agar Medium

Intended Use

Nutrient agar medium is used as a general purpose medium for the cultivation of less fastidious microorganisms, can be enriched with blood or other biological fluids.

Composition

Ingredients	Gms / Litre
Peptone	5.000
Sodium chloride	5.000
HM peptone B	1.500

Yeast extract	1.500
Agar	20.00
Final pH (at 25°C)	7.4 ± 0.2

Principle and Interpretation

Nutrient agar media are basic culture media used for maintaining microorganisms, cultivating fastidious organisms by enriching with serum or blood and are also used for purity checking prior to biochemical or serological testing. Nutrient agar medium is ideal for demonstration and teaching purposes where a more prolonged survival of cultures at ambient temperature is often required without risk of overgrowth that can occur with more nutritious substrate. This relatively simple formula has been retained and is still widely used in the microbiological examination of variety of materials and is also recommended by standard methods. It is one of the several non-selective media useful in routine cultivation of microorganisms. It can be used for the cultivation and enumeration of bacteria which are not particularly fastidious. Addition of different biological fluids such as horse or sheep blood, serum, egg yolk etc. makes it suitable for the cultivation of related fastidious organisms. Peptone, HM peptone B and yeast extract provide the necessary nitrogen compounds, carbon, vitamins and also some trace ingredients necessary for the growth of bacteria. Sodium chloride maintains the osmotic equilibrium of the medium.

OF Basal Medium

Intended Use

OF (Oxidation-Fermentation) basal medium is used for the determination of oxidative and fermentative metabolism of carbohydrates by Gram-negative bacteria.

Composition

Ingredients	Gms / Litre
Casein enzymic hydrolysate	2.000
Sodium chloride	5.000
Dipotassium phosphate	0.300
Bromothymol blue	0.080
Agar	2.000
Final pH (at 25°C)	6.8 ± 0.2

Principle and Interpretation

Hugh and Leifson developed OF medium to study oxidative and fermentative metabolism of carbohydrates by Gram-negative bacteria. This criterion is used during taxonomic studies of *Enterobacteriaceae*.

Casein enzymic hydrolysate in the medium provides the necessary carbon and nitrogen, vitamins etc required for bacterial growth. A carbohydrate whose fermentation reaction is to be studied is added separately. Phosphate buffers the medium and the low agar concentration determines motility and dispersion of the acid produced on the surface. Bromothymol blue acts as the pH indicator. The low concentration of agar permits the determination of motility and aids in the even distribution of any acid produced at the surface of the medium. Motility is observed as diffused zone of flaring out from the line of inoculation. Non-motile organisms grow along the line of inoculation.

Dextrose is the most important carbohydrate for use in OF basal medium. However, certain organisms may metabolize other carbohydrates even if they are unable to utilize dextrose. Degradation of the carbohydrate to acid is indicated by the pH indicator bromothymol blue which changes its colour to yellow. Oxidative utilization takes place when the medium is exposed to air while fermentative utilization occurs under exclusion of air.

The authors Hugh and Leifson showed that when a Gram-negative organism is inoculated in this medium containing a carbohydrate in duplicate, of which one tube is covered with mineral oil to exclude oxygen and the second tube is uncovered; reactions of differential value can be observed. Fermentative organisms will produce an acid reaction in both the covered and uncovered medium. Oxidative organisms will produce an acid reaction in the uncovered medium and gives light growth without change in the covered medium. Organisms which are not classified either as oxidative or fermentative show no change in the covered medium and an alkaline reaction in the uncovered medium.

Prepare the medium with 1% dextrose and without 1% dextrose. Two tubes of each carbohydrate are used per organism and inoculated by stabbing. One of the inoculated tubes of each carbohydrate medium is covered with 2ml of sterile mineral oil and the other is left uncovered. The tubes are incubated at 35-37°C for 18-48 hours or longer. The results are read after 48 hours.

The acidic reaction of oxidative organisms is more apparent at the surface of the medium that gradually spreads throughout the medium. If the oxidation reaction is weak or slow, an initial alkaline reaction at the surface of the uncovered tube may persist for several days and eventually convert to an acid reaction.

OF basal medium can be supplemented with 2% serum or yeast extract (0.1%) to make the medium more nutritious for the growth of bacteria.

Phenolphthalein Phosphate Broth

Intended Use

Phenolphthalein phosphate broth is recommended for identification of phosphatase positive *Staphylococcus aureus.*

Composition

Ingredients	Gms / Litre
Peptic digest of animal tissue	5.000
Beef extract	3.000
Sodium chloride	5.000
Sodium phenolphthalein phosphate	0.012
Final pH (at25°C)	7.4±0.2

Principle and Interpretation

Bacteria in the genus *Staphylococcus* are pathogens of man and other mammals. Traditionally they were divided into two groups on the basis of their ability to clot blood plasma (the coagulase reaction). The coagulase-positive *Staphylococci* constitute the most pathogenic species *Staphylococcus aureus*. The presence of *Staphylococci* in a lesion might first be suspected after examination of a direct Gram stain. However, small numbers of bacteria in blood preclude microscopic examination and require culturing first. Phosphatase has been implicated as a virulence factor for *Staphylococcus aureus*. The organisms produce both an acid and alkaline phosphates, the latter being repressed in the presence of inorganic phosphate in the medium.

Phenolphthalein phosphate agar is used for the identification of phosphatase-positive colonies of *Staphylococcus aureus*, which is a coagulase-positive pathogenic strain.

Peptic digest of animal tissue and beef extract supply the nitrogenous compounds, growth factors and trace ingredients essential for the growth of *Staphylococcus aureus*. Sodium phenolphthalein phosphate serves as a substrate for the phosphatase enzyme. Sodium chloride maintains osmotic equilibrium. Phosphatase production is determined by the liberation of phenolphthalein, which is indicated by the change in colour of the medium. When alkali is added to this medium, the liberated phenolphthalein gives a bright pink-red colouration. Alternatively phosphatase production can be determined by following technique.

Technique: Grow *staphylococci* overnight at 37°C on the medium. Invert the plate and pour few drops of ammonia solution in to the lid, read a positive culture whose colonies turn bright pink within a few minutes. The colour soon fades.

Photobacterium Broth

Intended Use

Photobacterium broth is used for cultivation and demonstration of luminescence of photobacteria.

Composition

Ingredients	Gms / Litre
Casein enzymic hydrolysate	5.000
Yeast extract	2.500
Sodium chloride	30.000
Ammonium chloride	0.300
Magnesium sulphate	0.300
Ferric chloride	0.010
Calcium carbonate	1.000
Monopotassium dihydrogen phosphate	3.000
Sodium glycerophosphate	23.500
Final pH (at 25°C)	7.0 ± 0.2

Principle and Interpretation

Most luminescent bacteria are marine; they are usually associated with fish and provide the light sources for fish luminescent organs and marking. The light-generating enzyme called luciferase is responsible for luminescence. Luminescent bacteria divert up to 10% of their respiratory enzyme into a specific metabolite pathway that converts chemical energy into visible light.

Photobacterium broth is prepared based on the formulations described by Daudoroff and Giese for the cultivation and demonstration of luminescence of photobacteria.

Casein enzymic hydrolysate and yeast extract provide nitrogenous compounds, carbon, sulphur, trace nutrients, vitamin B complex, which is essential for the growth of photobacteria. Monopotassium dihydrogen phosphate helps in maintaining the buffering capacity of the medium. Chlorides, sulphate, carbonate and glycerol phosphate helps for luminescence. The intensity of luminescence is related to the aeration of culture. The greater the oxygen supply, the greater will be the luminescence.

Pikovskaya's Agar

Intended Use

Pikovskaya's agar is recommended for detection of phosphate solubilizing soil microorganisms.

Composition

Ingredients	Gms / Litre
Yeast extract	0.500
Dextrose	10.000
Calcium phosphate	5.000

Ammonium sulphate	0.500
Potassium chloride	0.200
Magnesium sulphate	0.100
Manganese sulphate	0.0001
Ferrous sulphate	0.0001
Agar	20.000
Final pH (at 25°C)	7.0 ± 0.2

Principle and Interpretation

Phosphate exists in both organic as well as inorganic forms in soil. Organic matter derived from dead and decaying plant debris is rich in organic sources of phosphorus. However, plants are able to utilize phosphorus from soil only in the free available form. Soil phosphates are rendered available either by plant roots or by soil microorganisms. Therefore, phosphate-dissolving soil organisms play a part in correcting phosphorus deficiency of crop plants. Pikovskaya's agar was modified by Sundara Rao and Sinha for detection of phosphate-solubilizing bacteria from soil.

Yeast extract in the medium provides nitrogen and other nutrients necessary to support bacterial growth. Dextrose acts as an energy source. Different salts and yeast extract supports the growth of organisms. Phosphate-solubilizing bacteria will grow on this medium and form a clear zone around the colony, formed due to phosphate solubilization in the vicinity of the colony.

SIM Medium

Intended Use

Recommended for determination of hydrogen sulphide production, indole formation and motility of enteric bacilli from clinical and non-clinical samples.

Composition

Ingredients	**Gms / Litre**
HM Peptone B#	3.000
Peptone	30.000
Peptonized iron	0.200
Sodium thiosulphate	0.025
Agar	3.000
Final pH (at 25°C)	7.3 ± 0.2

Principle and Interpretation

SIM medium is used to differentiate enteric bacilli particularly *Salmonella* and *Shigella* on the basis of sulphide production, indole formation and motility. Jordan and Victorson reported that *Salmonella* paratyphi A and paratyphi B can be distinguished on the basis of H_2S production using lead acetate. Sulkin and Willett used triple sugar iron agar with 1% agar for motility along with H_2S production and carbohydrate fermentation. So asdescribed a peptone medium with low agar for motility and indole determination.

SIM medium enables determination of three characteristics by which enteric bacteria can be differentiated. Peptonized iron and sodium thiosulphate are the indicators of H_2S production. This H_2S reacts with peptonized iron to form black precipitate of ferrous sulphide. Motile organisms intensify the H_2S reaction. Motile organisms grow away from line of inoculation showing diffused growth while non-motile organisms grow along the stabline. Motility detection is possible due to the semi solid nature of the medium. Growth radiating out from the central stab line indicates that the test organism is motile. Peptone and HM peptone B provides nitrogenous and carbonaceous compounds, long chain amino acids, vitamins and other essential nutrients. Tryptophan from peptone is degraded by specific bacteria to produce indole. The indole is detected by the addition of chemical reagents following the incubation period.

Inoculate fresh culture with a single stab using straight needle through the center of the medium. Following incubation, observe for motility (diffuse growth outward from the stab line or turbidity throughout the medium) and for H_2S production (blackening of the medium). To detect indole production, add three or four drops of Kovac's reagent and observe for development of red color (positive reaction). Determine motility and H_2S production prior to determination of indole production.

Soil Extract Agar

Intended Use

Soil extract agar is used for isolation of soil microorganisms.

Composition

Ingredients	Gms / Litre
Glucose	1.000
Dipotassium phosphate	0.500
Soil extract	17.750
Agar	20.000
Final pH (at 25°C)	6.8 ± 0.2

Principle and Interpretation

The fertility of soil depends not only on the chemical composition but also on the qualitative and quantitative nature of microorganisms inhabiting it. Microorganisms inhabiting soil can be classified into bacteria, actinomycetes, fungi, algae and protozoa.

Soil is a natural medium for many organisms as they provide a perennial source of organic matter and other sources of carbon, nitrogen, minerals and vitamins required for their growth. Soil extract agar is a medium used for isolating soil microorganisms.

Soil extract provides all the essential nutrients required for growth of soil microorganisms. Glucose serves as readily metabolizable carbon source where as dipotassium phosphate buffers the medium.